Frederick Amrine

Seeing Ideas: Goethe's Science and Modernist Aesthetics

Studies on

Goethe's Science 2

"Was kann der Mensch im Leben mehr gewinnen,

Als daß sich Gott-Natur ihm offenbare?

Wie sie das Feste lässt zu Geist verrinnen,

Wie sie das Geisterzeugte fest bewahre."

Contents

It is a great paradox in the reception of Goethe that his theoretical writings on literature specifically, and art more generally, exerted little direct influence in the history of criticism, whereas his scientific writings would come to exert an important influence on the development of modernist aesthetics. Indeed, the influence of his scientific work is so pervasive that one might begin to wonder whether Goethe's much-derided late prediction to Eckermann that he would be remembered principally for his *Theory of Color* rather than his literary works might prove truer than anyone could have imagined.

In his own definitive history of color, John Gage has argued that Goethe's *Theory of Color* "has a claim to being the most important single text on colour for artists, and, indeed, for historians of art" [46]. It is well-known that Kandinsky was profoundly influenced by Goethe. As Britta Kaister-Schuster among others has argued, Goethe's *Theory of Color* figured prominently in the curriculum of the *Bauhaus*. Kandinsky's system of color was built upon the foundation of Goethe's primary polarity between yellow and blue, and Joannes Itten's deliberations on color were heavily influenced by Goethe's notion of a harmoniously balanced color circle, as were Paul Klee's theories of "color movement."

Goethe was also a major influence on the Russian Symbolist and anthrposophist Andrei Belyi (who wrote a lengthy epistemological treatise that is largely devoted to Goethe's scientific work), and of course Rudolf Steiner himself. Steiner edited Goethe's scientific writings for the great Weimar Edition, and wrote an essay promoting Goethe as the "Father of a New Aesthetics," but Steiner was also an important Expressionist architect, who chose the name "Goetheanum" for his two most important buildings. The second Goetheanum, which Wolfgang Pehnt has praised as "one of the most magnificent pieces of sculptural architecture of the 20th century" [148], and Hans Scharoun called "the most important architectural work of the first half of the twentieth century," still stands in Dornach, on the outskirts of Basel.

First Goetheanum

Second Goetheanum

According to Gage, the Expressionist Kirchner also turned to Goethe when he wanted to move on from Neo-Impressionism, and Paul Klee's writings on aesthetics borrow heavily from Goethe's plant and animal morphology. Anton von Webern founded his avant-garde *Theory of Tone* on Goethe's *Theory of Color*. as did his contemporary Hauer.

All these aspects of Goethe's reception by modern artists have been well covered in the secondary literature, and do not need to be rehearsed at greater length here. But we do need to account for this surprising affinity. Goethe and the aesthetics of Modernism would seem a paradoxical conjunction in any number of dimensions. After all, by the end of the nineteenth century, Goethe had come to represent the epitome of everything these artists despised: a smug *Bildungsbürgertum*, a reactionary Neoclassicism, the exclusionary hegemony of the worst form of canonicity. Why are these artists sleeping with the enemy? Why this strange attraction to Goethe?

One explanation, I think, is that, however inimical they may have felt towards Goethe as an artist, Goethe the *scientist* was as much of an outsider, as much of a revolutionary, as they were in the arts. After all, Goethe was in many ways an *avant-garde scientist*, anticipating some of the most important advances in the subsequent development of the philosophy of science. Even the anarchistic philosopher of science Paul Feyerabend has held up Goethe as a model, arguing that he had anticipated contemporary objections to the "dogmatic" treatment of Newton's theory of color. Goethe's warning that "[t]th most destructive prejudice is that which would put any branch of science under the ban" sounds rather like Feyerabend himself. Like many important recent philosophers of science, notably Thomas Kuhn, Goethe also doubted the possibility of "rational reconstruction" in science. Most important, Goethe was well aware that there is not, and cannot be, any neutral observation language:

> An extremely odd demand is often set forth but never met,
> even by those who make it: i.e., that empirical data should be
> presented without any theoretical context, leaving the reader,

the student, to his own devices in judging it. This demands seems odd because it is useless simply to look at something. Every act of looking turns into an observation, every act of observation into reflection, every active reflection into the making of associations; thus it is evident that we theorize every time we look carefully at the world ["Preface" to the *Theory of Color*].

Goethe was even present when they opening shot of the revolution in theory was fired by N. R. Hanson and his epoch-making chapter on the "theory-ladenness" of perception, which begins by quoting Goethe.

But there is an even more powerful explanation, I think. *I want to argue that, in some fundamental sense, Goethe's science is generally expressionist, and more specifically Cubist.* Such an assertion cries out for explanation, as I am well aware.

Please note that I am *not* attempting to argue that Goethe "influenced" Cézanne, Braque, and Picasso directly. Rather, is a matter of mutual clarification. Goethe can help us to understand Cubism, and Cubism (especially as interpreted and elaborated by David Hockney) can help us to understand better what Goethe was trying to accomplish in his morphology.

What Goethe was trying to convey is not at all obvious. Goethe struggled to explain what he meant, and even his eventual friend and closest collaborator Schiller struggled to understand, as Goethe himself records in his description of their first meeting, paradoxically entitled "Fortunate Encounter" [Goethe 18-21]. Bumping into each other at the end of a scientific meeting, Schiller and Goethe strike up a conversation. Surprisingly, we learn that the subject of conversation between these literary giants was not poetry or drama, but rather *botany*. Schiller asserts that the "fragmented way of dealing with nature" Offered by the Speaker be satisfactory, and Goethe is quick to agree, offering that "a different approach might well be discovered, not by concentrating on separate and isolated elements of nature, but by portraying it as alive and active, with its efforts directed from the whole to the parts" [Goethe 20].

Something astonishing happens next. At the climax of his argument, Goethe abandons words, and draws Schiller a picture. More surprising yet: as Goethe describes it, his drawing is *not* in the Neoclassical style that Goethe had just sought to master in Italy, even though this difference in taste was adduced specifically at the beginning of the essay as his excuse for avoiding Schiller. Rather, Goethe renders a "symbolic" sketch – a work of abstract art:

> We reached his house, and our conversation drew me in. There I gave an enthusiastic description of the metamorphosis of plants, *and with a few characteristic strokes of the pen I caused a symbolic plant to spring up before his eyes*. [Goethe 20]

For one fleeting moment, Goethe actually *becomes* an expressionist artist in order to convey to Schiller a precious *scientific* insight that he adjust attained in Italy while studying neoclassical art. This moment cuts a strangely attractive and paradoxical figure, both a parallel and a chiasmus: just as modern art rejects Goethe's aesthetics, but embraces his science, in a crucial moment, Goethe attempts to validate his science by developing an artistic practice that is the antithesis of everything he would seem to value as a visual artist.

Schiller tries earnestly to understand what Goethe means, but, ironically, the more revolutionary artist cannot interpret Goethe's abstract art:

> He heard and saw all this with great interest, with unmistakable power of comprehension. But when I stopped, he shook his head and said, "That is not an observation from experience. That is an idea." Taken aback and somewhat annoyed, I paused … I collected my wits, however, and replied: "Then I may rejoice that I have ideas without knowing it, and can even see them with my own eyes." [Goethe 20]

What is this strange *empirical expressionism*, this idea one can "see with [one's] own eyes?

Clearly, Goethe is anything but a Baconian empiricist or a naïve realist. But Goethe's morphology chromatics are not a "theory" or an "aesthetics" in the conventional sense of a conceptual structure, either. However, I *do* want to argue that scientific work is both a theory and aesthetics, and powerfully so, in the etymological sense of those terms. "Theory" comes from the Greek verb *theoreo*, which means "to see, to view," as is immediately apparent in "theory's" close relative "theater." (The word "idea" has undergone a similar transformation: it comes from the past participle of *eido*, and means literally "I-have-seen." If we restore the lost initial digamma, we see immediately its relationship to the Latin verb *video* and the English noun "video.") Goethe's science is also an *aesthetics* in the etymological sense of "perception" generally – and perhaps more specifically in the sense of Kant's "Transcendental Aesthetic," but pursuing that would lead us too far afield.

The crucial difference in Goethe's form of expressionism is evidenced already in the opening paragraphs of *The Metamorphosis of Plants*, which seems to begin conventionally enough, but conceals some startling methodological innovations:

1.

> Anyone who is paid even a little attention to plant growth will readily see that certain extended parts of the plant **frequently transform themselves** and **pass over into the form of the adjacent parts** – sometimes more, and sometimes less. [Miller 76; boldfaced passages re-translated by F.A.]

Here Goethe seems initially to be referring to the normal growth pattern of any annual plant, whereby the seed generates the cotyledons, the cotyledons the foliage these, etc. But this is not at all what he means, as becomes clear in the second paragraph, where he adduces certain *distortions* of the "normal" growth pattern, such as single flowers "doubling" when the stamens and anthers are displaced by pedals that "are either identical in form and color to the other petals of the corolla, or still bear *visible signs [sichtbare Zeichen]* of their origin." The ultimate development of this type is the proliferated rose or pink, in which entire small plants grow in place of the sexual organs, thus revealing the whole that is present, but usually invisible in each part.

Proliferated Rose (courtesy of Gordon Miller)

This seemingly pathological "backward step," as Goethe proceeds to describe it, "reversing the order of growth," ironically makes us "all the more aware of nature's regular course," with "the *law* of metamorphosis." This lawfulness lies hidden: in the paragraph immediately following, Goethe refers to it as a *"secret relationship'*[geheime Verwandtschaft], and Goethe uses similar language and other places. In paragraph seven, we are told that this *distorted* metamorphosis "will allow us to discover what is hidden in a leaf" [Goethe 77]; the fruitfulness of the plan is "hidden in a leaf" [Goethe 88]. It is not something immediately perceptible, and the Schiller might seem to have been right in claiming Goethe cannot be describing an empirical experience. But this hidden lawfulness of metamorphosis is a secret *hidden on the surface*; is one that is represented by *"visible signs"* [sichtbare Zeichen]. [...]

At the end of the essay, where the reader might expect a general theoretical statement, Goethe resists:

120.

> Here we would obviously need a general term to describe this organ which metamorphosed into such a variety of forms [note the subjunctive!] ... For the present, however, we must be satisfied with learning to relate these manifestations both forward and backward. Thus we can say that a stamen is a contracted petal or, with equal justification, that a petal is a stamen in a state of expansion; that is sepal is a contracted stem leaf with a certain degree of refinement, or that a stem leaf is a sepal expanded. ... [Goethe 96-97]

He resists substituting a static conceptual structure for an active experience of metamorphosis in all its fullness: such a structure would be, as Goethe calls it in the "Preface" to the *Theory of Color*, "the abstraction that we fear [Die Abstraktion, vor der wir uns fürchten] [Goethe 159[1]]. But this is not an appeal to naïve realism, either. Rather, Goethe calls on us to perform a powerful transformation of the scientific vision in another sense. He calls on us to overcome the single perspective of naïve realism by practicing looking at the plant from multiple perspectives, by *increasing* our perceptual mobility, running through the forms forwards, backwards, constantly relativizing the spatial frames surrounding each individual form. The ideality of the archetype that we learn to perceive directly is not a function of its remoteness *from* the phenomena, but rather of our increased perceptual mobility and engagement *with* the phenomena. Goethe's scientific aesthetic strives to attain not an abstract hypothesis, a propositional statement about "plantness," but rather a kind of *intimate seeing*, and Goethe's *Metamorphosis of Plants* is a workbook that invites us to

---

[1] Unfortunately, Miller's translation, "the pitfalls of abstraction," takes much of the edge off this famous phrase.

12

engage in specific practices to the end of developing a kind of artistic faculty.

Properly understood and employed, Goethe's *Metamorphosis of Plants* is itself a workbook adequate to this end, but others have created even more powerful workbooks in the spirit of Goethe. My own essay in the Reidel volume that Hans Zucker and I co-edited seeks to describe in some detail a scientific study by Jochen Bockemühl. That study is much too complex to recapitulate in detail here, but suffice it to say that Bockemühl's botanical research leads us through ever more complex suites of foliage, progressing from the embryological development of a single leaf [Fig. 1]

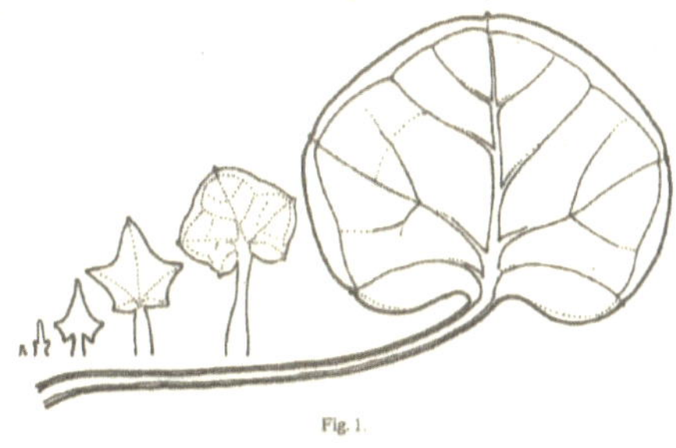

Fig. 1.

to the complete sequence of foliage leaves [Fig. 4]

Fig. 4.

13

Here is one of Cézanne's many paintings of Mont St. Victoire:

And here is a photograph of the same motif:

Note how much larger the mountain looms in the painting than in the photograph.

This is one of several devices that Cézanne generally employs "to reverse scientific perspective" by "containing space close to the canvas plane" [Machotka 23]:

- slowing down the recession of a curve;
- assimilating receiving diagonals to the horizon;
- enlarging distant objects or raising the horizon; and
- tilting planes upwards. [Machotka 23-24]

(Apparently, Cézanne accomplished the last of these by, for example, propping up plates in still-lifes with coins!) Cézanne's perspectival "distortions" are still mild, however, compared even with Picasso's early Cubist phase –

Picasso, "Houses on the Hill" (1909)

– or for example Picasso's painting "Fan, Box, Salt, Melon" of 1909:

How are we to understand Cézanne's seeming reversal of "scientific perspective," and the radical spatial "distortions" that seem to underlie the project of Analytic Cubism? I would argue that we can go a long way towards understanding Cézanne's spatial devices by invoking an effect well-known and widely discussed in the psychology of perception called "constancy scaling." One simple experiment that reveals this effect is to ask someone to stand 10 paces away from you in a space without too many other visual clues, and then to ask another of the same height to stand 20 paces away. Although twice as distant, and thus projecting a retinal image half as large, the person further away invariably appears two-thirds to three- quarters as large. We do not just interpret, but *actually see* things as closer to their real, psychological proportions, despite the sensory data to the contrary. Another simple demonstration is to hold a dime or a quarter in each hand at eye level, and then to adjust them so that one appears half the size of the other.

Most perceivers find themselves holding the coin that is meant to appear smaller much more than twice as far away.

In constancy scaling, our *inner activity* overrides the retinal image; we create a new perspective *within* the phenomena as a way of seeing phenomena *in the way we know inwardly they really are*. Psychology – which is to say, *expression* – trumps the empirically given. A seeming sensory illusion is created to the end of a greater *psychological realism*.

As in *The Metamorphosis of Plants*, Goethe begins his *Theory of Color* "pathologically," invoking what had previously been dismissed as a set of subjective "illusions" as the foundation for his own rigorous study of color. He begins with the phenomena he calls "Physiological Colors," which have since come to be called simultaneous and successive contrast. There is a striking demonstration of this effect in a Danish documentary on Goethe's color theory by Marie-Luise Lefèvre, Marie-Luise Lauridsen, and Hendrik Boëtius.[2] The colored shadow is always seen in the complementary color, *even though it has no measurable wavelength*. In some sense, we see a color that "is not there." At every moment, the mind actively produces the complement of the empirically perceived color, completing the color circle and thereby restoring harmony through wholeness. And yet, Goethe argues, this phenomenon is not merely subjective: colored shadows behave lawfully, and the law governing their generation has full predictive value. As in constancy scaling, a seeming illusion opens a window on the fundamental *law* of perception: *the perceiver is actively involved in*

---

[2] Available on YouTube at
https://www.youtube.com/watch?v=HtMvYHq2-uI&list=PLLEMpXvr0s1TrnjwOMG7__JWpayNr8tSf
See the especially the segment from minute 2 to minute 4. The film can be purchased from

Magic Hour Films ApS
Fortunvej 56
DK-2920 Charlottenlund
+45 3964 2284
+45 3582 1333
www.magichourfilms.dk
post@magichourfilms.dk

constructing experience. Such "archetypal phenomena," as Goethe calls them, are especially revealing because they are not merely *products* of perception: instead, they represent immediately the *process* of perceptual *activity*.

The seeming distortions in Cézanne's late paintings need to be seen in the same way: as an attempt *to paint the process of perceiving together with its product*. I propose that Analytic Cubism should be viewed in the same light. In the same way that Bockemühl's contemporary research illuminates Goethe's, we can gain valuable insights into Analytic Cubism by studying the work of a later artist who was inspired by the Cubists, and tried in his own way to carry their work further.

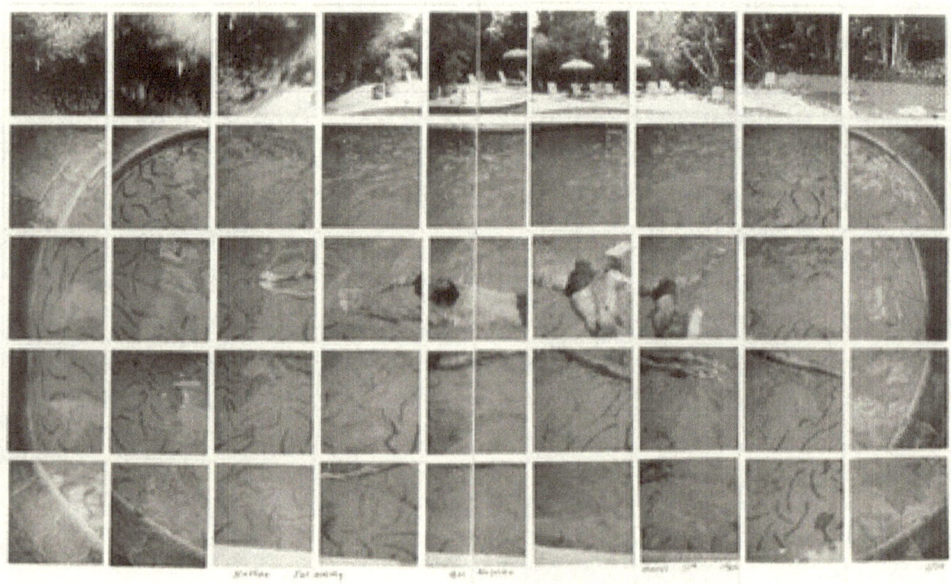

David Hockney's remarkable photo collages are collected in a volume entitled *Cameraworks*. This portrait of his friend swimming is on the book jacket. Hockney's collages deliberately recall the Cubists –

Indeed, some are even self-ironic in the overtness of their allusiveness (as in the following still-life), while others are more subtly original.

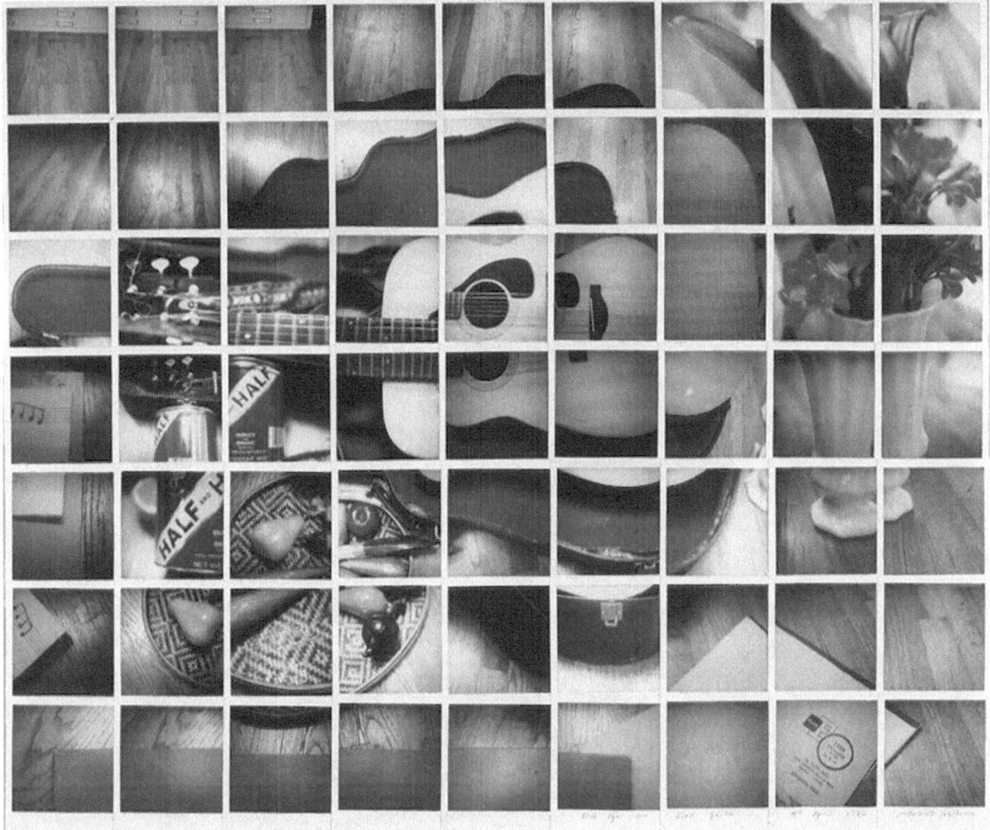

As he pursued this project, Hockney eventually came to realize that the square borders of the Polaroids he had used initially were perpetuating the alienated "window view" of Renaissance single-point perspective that both he and the Cubists had been so eager to dismantle. Thus he began composing with borderless 35mm photos (despite the much higher degree of difficulty), with stunning results. Cézanne had pulled the objects he painted as close as possible to the surface of the canvas, trying to reduce as much as possible the distance between the observer and the object, but a Hockney goes even further, projecting the

observer right into the picture, reducing the distance to *zero*: the brown wingtips at the very bottom are Hockney's own shoes.

When these photo collages were first displayed in 1988, Lawrence Wechsler interviewed Hockney, and then publish the interview in *The New Yorker* under the title "True to Life." (A slightly longer version appeared later as the preface to *Cameraworks*.) Hockney has lots to say about Cubism in this interview, much of it genuinely philosophical. Most important for our purposes, Hockney is profoundly illuminating on the subject of the Cubists' supposed abstraction. Turning the conventional wisdom on its head, Hockney argues that it was not the camera, but the Cubists who were more "true to life." The Cubists' seemingly abstract, formalistic technique of building up an object by representing it for multiple perspectives within the same picture; of conveying the sensation of having walked around their objects, of a space that "enveloped and flowed over and around the objects in the picture" [Waddington 19], is recapitulated in Hockney's ever-varying camera angles and focal lengths. The "general perspective" is built up from myriad "micro-perspectives" [Wechsler 63]. The objectifying triangulation of bifocal vision dissolves as the distance between perceiver and perceived disappears; *we* are projected onto the plane of the canvas, and objects begin to swim, like Picasso's portraits of his lover.

Thus Cubism is anything but a dry formalism that dissolves objects into intersecting planes. Rather, Cubism is an *aesthetic practice* that leads to "a kind of intimate seeing" [Wechsler 64]. Or, as Hockney himself puts it:

> Analytic Cubism in particular was about perception – about the difficulty of perception. … I've recently been reading a lot of books about Cubism, and I keep coming upon discussions of intersecting planes, and so forth, as if Cubism were about the structure of the *object*. But really it's rather about *the structure of seeing* the object. [64]

Like Goethe's scientific works and Cubist painting, Hockney's collages are *workbooks* in which we can practice *enhanced perception generally*: Wechsler calls them "a school for vision" [63]. Studying the Cubists, and working on his photo collages, Hockney came to understand vision as a dynamic, mobile, almost *tactile* process, "a continuous accumulation of details perceived across time and synthesized into a larger, continuously metamorphosing whole" [Wechsler 63]. The problem with conventional photography, according to Hockney, is that it turns us into a "paralyzed Cyclops":

> But, more important, I realized that this sort of picture came closer to how we actually see – which is to say, not all at once but in discrete, separate glimpses, which we then build up into our continuous experience of the world. Looking at you now, my eye doesn't capture you in your entirety. I see you quickly, in nervous little glances. I look at your shoulder, and then at your ear, your eyes (maybe, for a moment, if I know you well and have come to trust you – but even then only for a moment), your cheek, your shirt button, your shoes, your hair, your eyes again, your nose and mouth. There are hundred separate looks across time from which I synthesize my living impression of you. And this is wonderful. If, instead, I caught all of you in one frozen look, the experience would be dead. It would be like – like looking at an ordinary photograph. [Wechsler 62]

According to Hockney, the Cubists understood this dynamism in perception, and tried to paint *the process of seeing* in all its intimacy and immediacy, but, unfortunately, the Cubists' project was misunderstood, and led to an abstraction that was merely formal. It led not to the "intimate seeing" of psychological realism, but to the alienating formalism of abstract expressionism:

> The great misinterpretation of twentieth-century art is the claim advanced by many people, especially critics, that Cubism of necessity led to abstraction – that Cubism's only

true heritage was this increasing tendency toward a more and more insular abstraction. But, on the contrary, Cubism was about the real world. It was an attempt to reclaim territory for figuration, for depiction. Faced with the claim that photography had made figurative painting obsolete, the Cubists performed an exquisite critique of photography. The show that there were certain aspects of looking – basically the human reality of perception – that photography couldn't convey, and that you still needed the painters hand and eye to convey them. [Wechsler 70]

The strange thing about the legacy of Cubism is that it did exert an influence on abstract artists during the 'thirties and 'forties but was virtually ignored by the realists. … I think that now – as the cul-de-sac of abstraction has become increasingly self-evident – for painting to go forward we have to go back to Cubism. It still has very important lessons to teach us. [Wechler 71]

In a similar vein, I would contend that we still have much to learn from *Goethe the scientist*. We can imagine an alternative path that science might have taken as well, a path on which phenomena are not divorced from the observer; qualities are not relegated to secondary, subjective status, but investigated with aesthetic rigor; where reason is not instrumentalized, but reconciled with affect, and guided to humane ends. For the past 200 years, Goethe has been, and for the foreseeable future he will remain, the inspiration and best hope for the development of such an alternative.

Bibliography

Amrine, Frederick, Francis J. Zucker, and Harvey Wheeler. *Goethe and the Sciences: A Reappraisal*. Dordrecht: D. Reidel, 1987.

Amrine, Frederick. "Goethean Method in the Work of Jochen Bockemühl." In *Goethe and the Sciences*, pp. 301-318.

Belyi, Andrei. [Bugaev, B.] *Rudol'f Shteiner v mirovozzrenim sovermennosti*. Moskva: Dukhovnoe znanie, 1917.

Feyerabend, Paul. "Classical Empiricism," in *The Methodological Heritage of Newton*, ed. Robert E. Butts and John W. Davis (Toronto: Univ. of Toronto Press, 1970), pp. 150-70.

Gage, John. *Colour and Meaning: Art, Science, and Symbolism*. London: Thames and Hudson, 2000.

Goethe, J. W. von. *Scientific Studies*. Ed. and trans. Douglas Miller. Vol. 12 of *Goethe Edition*. New York: Suhrkamp Publishers, 1988.

Hanson, Norwood Russell. *Patterns of Discovery: An Inquiry into the Conceptual Foundations of Science*. Cambridge: Cambridge University Press, 1958.

Hockney, David. *Cameraworks*. New York: Knopf, 1984.

Kuhn, Thomas. *The Structure of Scientific Revolutions*. 50th Anniversary Edition. Chicago: University of Chicago Press, 2012.

Loran, Erle. *Cézanne's Composition: Analysis of his Form with Diagrams and Photographs of his Motifs*. Berkeley: University of California Press, 1950.

Machotka, Pavel. *Landscape into Art*. New Haven: Yale University Press, 1996.

Pehnt, Wolfgang. "The Architecture of Rudolf Steiner." In his *Expressionist Architecture*. London: Thames and Hudson, 1973, pp. 137-148.

Steiner, Rudolf. "Goethe als Vater einer neuen Aesthetik." In his *Kunst und Kunsterkenntnis: Grundlagen einer neuen Aesthetik.* Dornach: Rudolf Steiner Verlag, 1985, pp. 13-36.

Waddington, C. H. *Behind Appearance: A study of the relations between painting and the natural sciences in this century.* Edinburgh: Edinburgh University Press, 1969.

Wechsler, Lawrence. "True to Life." *The New Yorker,* July 9, 1984, pp. 60-71.